I0709726

curiousabout
COMPOSTING
BY AMY S. HANSEN
AMICUS LEARNING

What are you

curious about?

Curious About is published
by Amicus Learning,
an imprint of Amicus.
P.O. Box 227
Mankato, MN 56002
www.amicuspublishing.us

Editor: Alissa Thielges
Series Designer: Kathleen Petelinsek
Book Designer and Photo Researcher: Emily Dietz

Library of Congress Cataloging-in-Publication Data
Names: Hansen, Amy, author.
Title: Curious about composting / by Amy S. Hansen.
Other titles: Composting
Description: Mankato, MN: Amicus Learning, [2025] |
Series: Curious about green living | Includes bibliographical
references and index. | Audience: Ages 5–9 | Audience:
Grades 2–3 | Summary: "What is composting and how
does it work? Ignite kids' growing curiosity about the green
living with an inquiry-based approach to composting. Includes
infographics and back matter to support research skills, plus
table of contents, glossary, and index"—Provided by publisher.
Identifiers: LCCN 2023043270 (print) | LCCN
2023043271 (ebook) | ISBN 9781645496960
(library binding) | ISBN 9781681529653
(paperback) | ISBN 9781645497028 (ebook)
Subjects: LCSH: Compost—Juvenile literature.
Classification: LCC TD796.5 .H359 2025 (print) | LCC
TD796.5 (ebook) | DDC 631.8/75—dc23/eng/20231115
LC record available at https://lccn.loc.gov/2023043270
LC ebook record available at https://lccn.loc.gov/2023043271

Photo Credits: Alamy: Janet Horton, 18–19; Freep!k:
nastiklis1992, 2 (l), 8; iStock: Antonio_Diaz, 21, domnicky,
11 (b), Marizza, 3, 17, SDI Productions, 20, svetikd, 16,
VectorMine, 6–7; Noun Project: Ken Murray, 22 & 23 (apple),
P Thanga Vignesh, 22 & 23 (worm); Shutterstock: brgfx, 13,
Brian Simmonds 14, Elena Wein, cover, 1, Natalliaskn, 4–5,
OMfotovideocontent, 7, Oscar Martinez Troncoso, 11 (t),
Petro Perutskyi, 12, svetlatek, 9, wikkie, 2 (r), 15

Printed in China

What is compost?

It's slop! Drop an apple core into the **compost** bin. It sounds squishy. The bin has food scraps and grass clippings in it. It might have leaves and sticks, too. Composting recycles! Everything in the bin **decays**. It becomes fresh soil.

People collect kitchen scraps to add to a bigger compost pile.

How does old food turn into soil?

The apple core turns brown in the air. Then worms and insects tear it apart. Next, tiny **bacteria** in the soil **digest** it. The apple gets slimy. Soon the apple decays. There is only rich soil left in the bin.

DID YOU KNOW?
It may take a year for apple pieces to turn into fresh soil.

Food scraps turn into valuable soil in a compost bin.

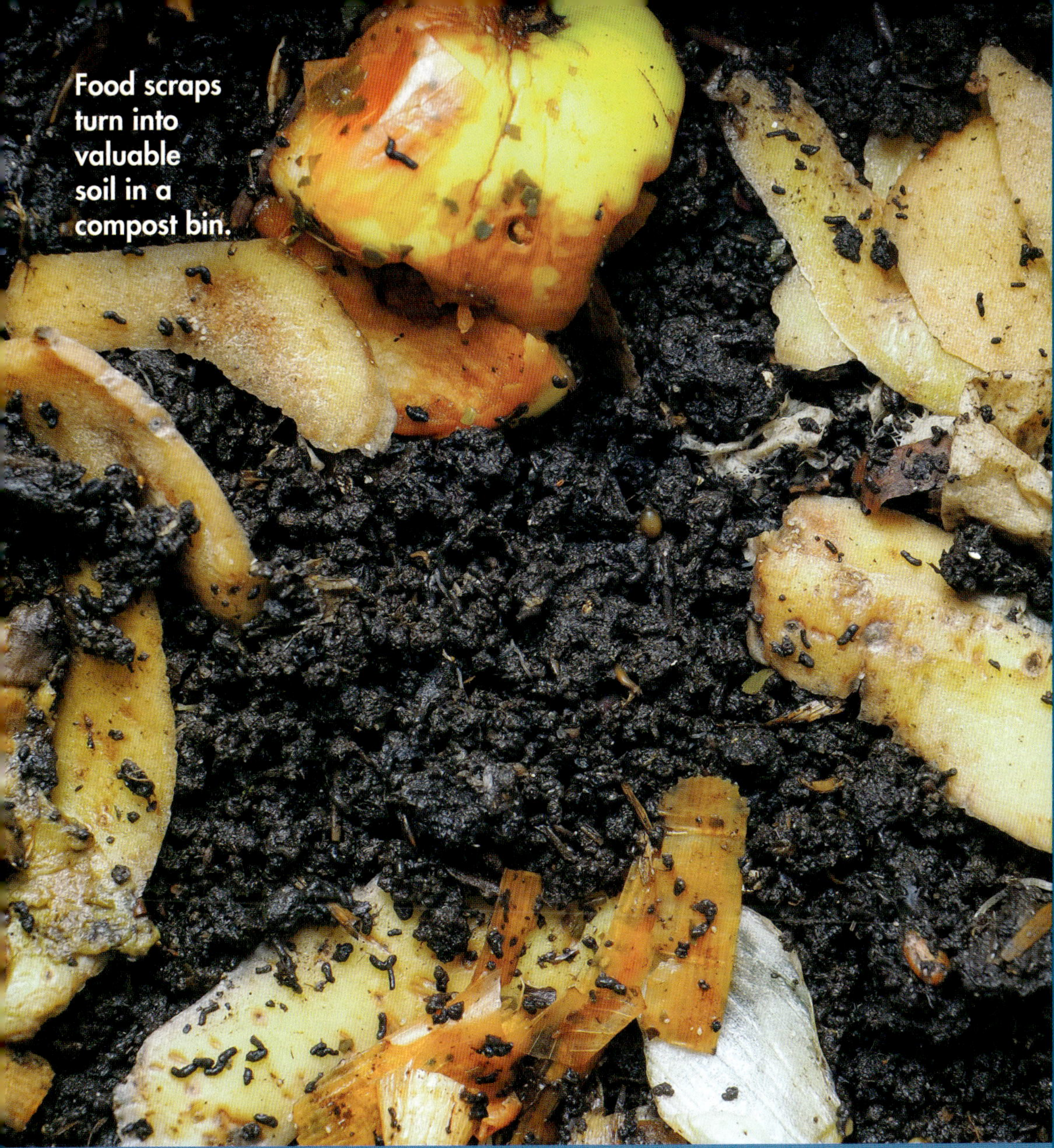

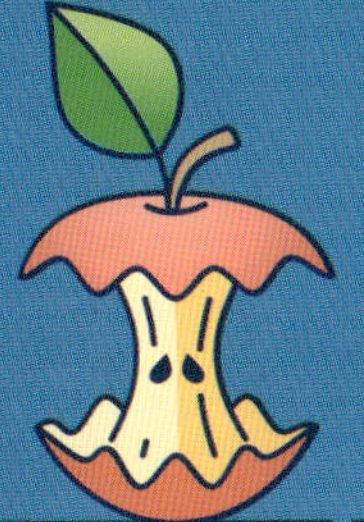

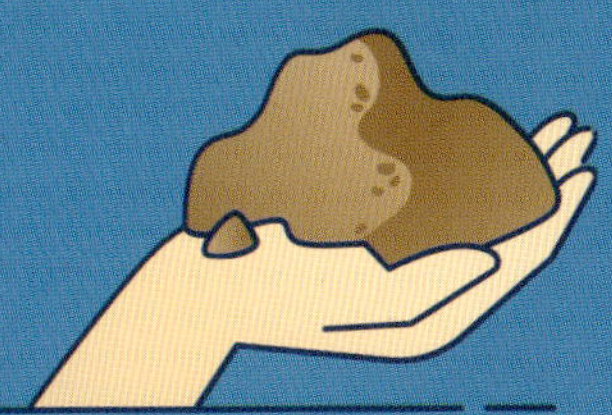

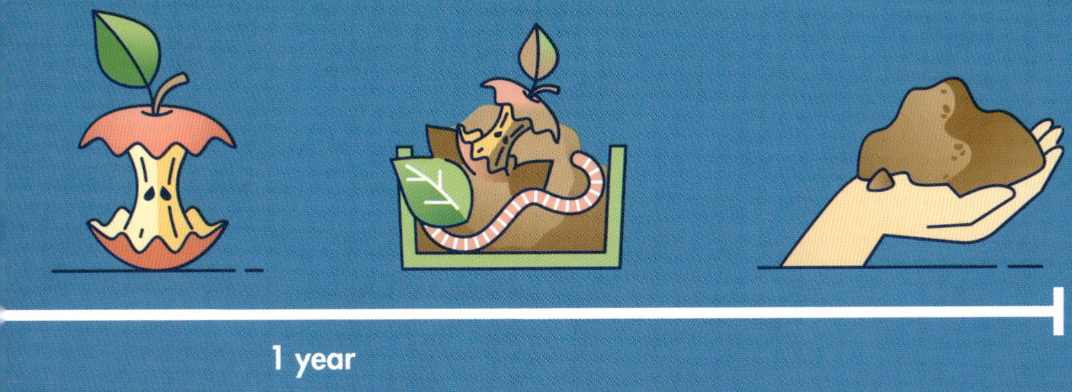

What can I put in compost bins?

Anything that can rot! There are two ways to compost. A hot compost is hot in the middle. You can throw meat, fish, dairy, and cooked scraps in this bin. The pile decays in a few weeks. A cold compost can take a year. This bin needs veggie scraps and dead leaves.

HOT VS. COLD COMPOST

HOT COMPOST
3 WEEKS TO 3 MONTHS

Cooked Food

Oils

above 115 °F
(46 °C)

Cheese

Bones

COLD COMPOST
1 TO 2 YEARS

Apple Core

Eggshells

Banana Peel

below 100 °F
(38 °C)

Shredded
Newspaper

Dry Leaves

Does compost stink?

It shouldn't! The stink means it's too wet. Grab a shovel and turn the pile upside down. Then air will reach the wet parts. A pile may really stink if it has too many "greens." These are veggie, fruit, and plant scraps. Balance your pile with more "browns." Mix in dry leaves, newspaper, or cardboard. These soak up moisture and help air get in the pile.

You can use a
rake to help turn
your compost.

Worms help break
down food and
animal waste.

DID
YOU KNOW?
One kind of composting
uses worms. It is called
vermicomposting.

Can composting help the Earth?

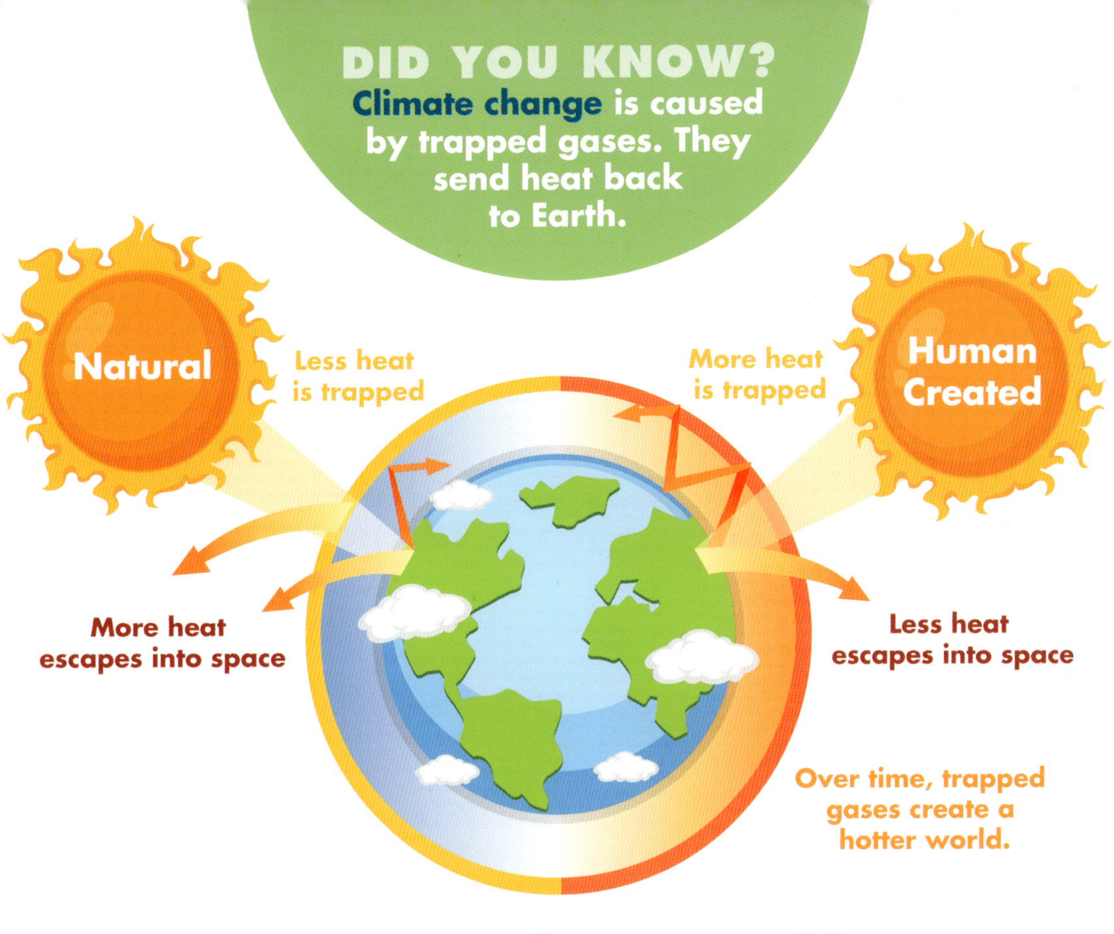

The Greenhouse Effect

Yes! When food decays, it gives off a gas. In landfills, this gas is trapped under all the garbage. This creates **methane** gas. Compost bins create a gas called **carbon dioxide**. Both are **greenhouse gases**. But methane is much worse. It traps 30 times more heat in the air. Reducing this gas will help slow climate change.

Food waste in a landfill does not help the Earth.

What happens if we don't compost?

We will miss out on valuable soil! Food scraps will decay anywhere. If they end up in a landfill, they don't get reused. Plus, landfills are running out of space. Composting gives us a soil called **humus**. It is full of **nutrients**. Plants grow easily in this.

Compost soil helps
plants grow big
and strong.

How do I compost at home?

After cutting up veggies, you can add the scraps to a bucket.

It's easy! Get a small bin for the kitchen. When you chop fruit and veggies, put the scraps in the bin. Then add a "brown" layer to prevent a stink. When you cook eggs, you can put in the eggshells. Keep adding "green" and "brown" layers. When the bin is full, dump it into a larger compost bin outside.

Two kids dump scraps into an outdoor compost bin.

What if my home doesn't have room to compost?

Some cities have composts near community gardens.

Check around your community. Many cities have a compost center. Workers set up a big bin in a parking lot. They may use hot compost systems. Then they can take many food scraps. There are also systems for just paper, wood, and leaves. Cities may give the compost soil to gardeners.

DID YOU KNOW?
In the United States, about a third of what we throw away could be composted. That's a lot of waste!

Can we do anything to help?

Yes! Students can set up compost bins at school. Some schools start gardens with the rich soil. Others use it to reduce their waste. Some students send letters to their state's governor. These students say every school should compost. It is one way we can help stop climate change.

ASK MORE QUESTIONS

How do you know when compost is ready to use?

Why does a hot compost break down into soil faster?

Try a BIG QUESTION: What role does compost play in nature?

SEARCH FOR ANSWERS

Search the library catalog or the Internet.
A librarian, teacher, or parent can help you.

Using Keywords
Find the looking glass.

Keywords are the most important words in your question.

If you want to know about:

- when compost is ready to use, type: WHEN IS COMPOST DONE

- how hot compost works, type: HOT COMPOST PROCESS

FIND GOOD SOURCES

Are the sources good?
Some are better than others.
An adult can help you. Here are
some good, safe sources.

Books

Break It Down: A Look at Composting
by Heather DiLorenzo Williams, 2023.

The Organic Lifestyle
by James Shoals, 2020.

Internet Sites

Make a Composter!
*https://static.pbslearningmedia.org/media/
media_files/composter.pdf*
PBS is a public TV network. It creates
kid-friendly educational shows and activities.

Make the Most of Compost!
*https://www.youtube.com/
watch?v=Q5s4n9r-JGU*
SciShow Kids is a science YouTube channel
for young, curious minds. It is run by a team
of scientists who fact check every detail.

Every effort has been made to ensure that these
websites are appropriate for children. However,
because of the nature of the Internet, it is impossible
to guarantee that these sites will remain active
indefinitely or that their contents will not be altered.

SHARE AND TAKE ACTION

Collect kitchen scraps in a bucket.
When it is full, dump it in
a community compost.

Create a mini compost in a glass jar.
You'll have nutrient soil in a few
months, and you get to watch it
decay right before your eyes!

Grow a garden.
Test to see how plants grow
with and without compost.
Which plants grow better?

GLOSSARY

bacteria Tiny life forms that live all over the world. Some cause diseases but most do important jobs, like breaking down food.

carbon dioxide A colorless gas that is formed in the process of breathing.

climate change Changes in the Earth's weather patterns.

compost The process of breaking down food, leaves and wood.

decay To rot away.

digest To break down food so it can be used for energy.

greenhouse gas A gas in Earth's atmosphere that traps heat.

humus A dark material that forms when plants decay.

methane A colorless gas that has no smell and that can be burned for fuel.

nutrient A substance or ingredient a person, animal, or plant needs to be healthy.

INDEX

About the Author

Amy S. Hansen lives in Maryland. She started writing as a reporter for local newspapers. After a few years she went back to school to study environmental science and then worked as a science reporter. Nowadays, she (mostly) writes for kids because it is much more fun.